PREFACE.

THE scope of this little book may be very briefly yet very clearly stated, both positively and negatively. It is intended for those students of electrical science who have some elementary knowledge of the theory and who desire to make a flight into the practical work of designing and building a small dynamo or motor. It will at the same time serve the ends of those whose attempts in this direction have not been wholly satisfactory, and is calculated to help them both to improve on their earlier attempts and to construct more ambitious machines. The present work will present no difficulty to an amateur of average intelligence who takes a lively interest in the subject and has some slight grasp of draughtsmanship in addition to constructive talent. Plain diagrams, drawn to scale, are here given of numerous types of dynamos. A system has been adopted, it is believed for the first time, by which a dynamo or motor of any given size within limits can be built from any of the drawings, and the reader is assured that any machine made correctly from the instructions herein provided will prove entirely satisfactory in its work.

F. E. POWELL.

368543

SMALL DYNAMOS AND MOTORS.

CHAPTER I.

GENERAL CONSIDERATIONS FOR THE AMATEUR DYNAMO OR MOTOR BUILDER.

THE attempt made in this little work to present in compact form a large variety of dynamos and motors in "standard" dimensions and with tabulated windings for different outputs may be frustrated by very trifling departures in the mode of construction or by the alteration of even minor details by the inexperienced builder. It should be said, however, that this possibility is not by any means the fault of the scheme, since it is shared by every work, great or small, ever published on the subject. It is necessary to guard the reader against failure by stating plainly the conditions under which the following dynamos and motors will work, and to this very important end the remainder of Chapter I is consequently devoted.

Probably every reader fully understands the necessity for good workmanship in even the smallest dynamo.

Bad workmanship undoubtedly results in inefficient working, whether the fault lies in the important parts of a field magnet or armature, or in such minor details as the bearings or the driving mechanism. On the other hand, however good the workmanship no machine will give its stated output continuously and safely unless the constructive materials are fully up to the standard kept in view when designing it. For the guidance of the practical amateur the following details are enumerated under this head.

The field magnets are to be in soft grey cast iron except in the special instances and for parts of certain types, where soft iron forgings or rod are specified; joints to be accurately faced up or machined; armature tunnel accurately cored or bored to size.

The armature stampings to be of best Swedish iron, and to be painted or varnished on one side.

The wire used throughout must be of the very best quality, and should be new. Old wire is often not only hard and brittle, but its conducting power is lessened, and the presence of kinks, etc., renders it impossible to wind so much in a given space. The quantities given in the tables must be adhered to, and the windings must of course be exactly as stated. The wire should also be of good quality, as cheap wire is insulated with thick and uneven covering, and is usually of poor conductivity.

Under the above conditions a machine made from the data in the following pages can be safely depended upon to give satisfactory results, and an output of electrical or mechanical energy in accordance with its specified rating.

Dynamos and electro-motors of the types illustrated in the following pages can be employed for either purpose indifferently, a statement which also applies almost without qualification to the larger machines these models represent. The chief difference of any importance to the amateur is in the question of weight, and since the employment of cast iron for the field magnets implies almost double the weight for the finished machines over those in which soft forged iron is used, the reader who desires to build a *light* motor (for use in a model electric launch, for example) should make the field magnets of soft wrought iron bar or of Swedish iron stampings. He may take it that quite half the weight of the field magnets may thus be saved without impairing the efficiency of his machine; it is not apprehended that the reader, of average skill, will find any difficulty in thus calculating the size of the field magnets required, since all the data for the ordinary machine in which cast iron is employed are given in the following chapters.

The use of wrought iron in the field magnets will reduce the amount of wire required in the exciting coils, since it is not the *length* of wire, but the number of turns which determines the degree of magnetisation. In the wiring tables in Chapter V the amount of wire set apart for any field magnet exciting coil is calculated for cast iron magnets, but, generally speaking, about three-quarters of this amount would produce an equivalent number of turns on a field magnet of wrought iron of half the sectional area. The employ-ment of a shorter wire, however, involves a less re-sistance, in consequence of which a larger current would flow. This would waste current, and the difficulty may have to be met by using a size smaller wire, taking care to keep as nearly as possible to the same resistance and number of turns. These remarks apply, whether the machine is intended for dynamo or motor.

Field magnets for *dynamos* if made of iron of too soft a character, such as Swedish iron stampings, often are unable to retain sufficient residual magnetism to excite the machine at starting. Stampings for field magnets should therefore be used only for motors, in which case they have every advantage over cast iron magnets.

Patterns for small dynamos should be very carefully made. The reader is strongly advised to have his castings cored for the armature tunnel, as, if this is properly done, there will be no need to bore out the space, and a tedious operation is thus omitted. It is needless to remark that if this plan is to be adopted, the finished diameter of the armature core must be known. First obtain the stampings, and the diameter of the cored hole can then be determined, as well as the chief dimension (diameter) of the armature spindle.

Joints in the field magnet are in every way objection-able. When the design permits of no other course, or the special advantages of a partly wrought and partly cast magnet are desired, the joints should be as perfect as possible, to diminish magnetic resistance. Practi-cally for the same reason the "air-gap" surrounding the armature core should be reduced to a minimum, and this is a point requiring special care. The gap between the outer surface of the armature and the inner surface of the tunnel should never be less than a bare ⅟₃₂ inch, or there is danger of abrasion. For the smallest armatures ⅟₃₂ inch is a fair amount, and for the largest of those described in these pages ⅟₃₂ inch may be allowed.

Small motors should generally be "series" wound. Self-regulation, which is a feature of the shunt machine, is not of any considerable importance in a very small motor, and the series-wound machine possesses the very great advantage of a large "starting torque," which in other words means that it is able to start under a much greater load than a shunt-wound machine.

Ordinary bi-polar dynamos and motors are wound so as to produce one north and one south pole. The winding of a single-coil magnet, such as that shown in fig. 5, is obviously simple enough, as it is continuous and in one direction only. The proper method of winding any magnet of the general form shown in fig. 8 is here indicated, the thick lines representing the turns of wire (fig. 1). Fig. 1A is a diagram of "shunt" winding, and fig. 1B shows the proper connections for "series" winding. Magnets of the "Manchester" type (fig. 12) or of the form shown in fig. 15 are wound to produce "consequent" poles, "north at top and south beneath," or *vice versa.*

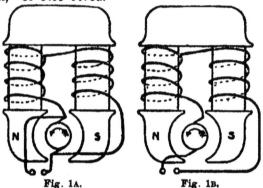

Fig. 1A.   Fig. 1B.

Methods of Winding an Ordinary Dynamo Magnet having Two Cores.

Correct diagrams for winding each of these forms are given in figs. 2 and 3 respectively. The usual method of winding a four-pole machine is such as to

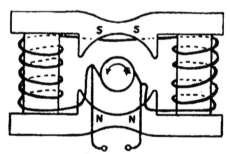

Fig. 2.—Diagram showing Winding of Manchester-type Field Magnets.

produce alternate N and S poles. This is indicated in fig. 4.

Before a new dynamo can be expected to "work," it must have the field magnets magnetised in order to provide the small amount of residual magnetism (which will be permanently retained by the iron) necessary to render the machine self-exciting. This magnetisation must be of correct polarity, which can be determined by means of a small pocket compass. The best source of current for the purpose is a good bi-

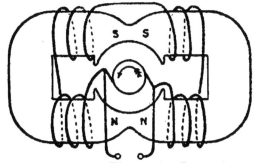

Fig. 3—Diagram showing Winding of a "Four-Coil" Magnet.

chromate battery or accumulator, of as high a voltage as possible. The proper polarity can be determined from a knowledge of the winding, although for small dynamos it is usual to magnetise the fields in one direction, run the machine, and observe whether it "builds up," that is, excites itself, or not. If not, the polarity is reversed by reversing the connections of the exciting coils to the battery.

For charging accumulators, certain dynamos are inadmissible as the source of current. For this purpose a machine *must* be shunt-wound, and must not have a

Siemen's (or H) armature. A dynamo with tri-polar
or quadri-polar armature will serve, but a much better

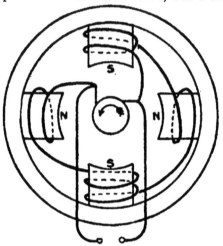

Fig. 4.—Diagram of Winding for a Four-Pole Machine.

plan is to have a shunt-wound drum-type machine with
as large a number of sections as possible. Compound-
ing is not recommended for small machines.

CHAPTER II.

FIELD MAGNETS: TYPES AND SIZES FOR VARIOUS OUTPUTS.

THE field-magnet designs shown in this chapter are drawn to scale and represent the best proportions of each of the well-known types. A few words of explanation will enable the reader to make good use of these diagrams.

This series of outline drawings is based primarily on the number of exciting coils with which the magnets are furnished. Three distinct groups are thus obtained, although the second is again subdivided into magnets with "consequent poles" (Manchester-type, for example), and the ordinary "under" and "overtype" bi-polar machines.

It will be noticed that the three diagrams on page 14 (figs. 5, 6 and 7), are to all intents copies of one another, the only differences being in the supports or feet. It is recommended that machines of this type be cast in one piece, as this secures the greatest efficiency, and the cores are not difficult to wind. If the reader, however, is desirous of constructing a machine with wrought iron core as at A, fig. 5, this can be

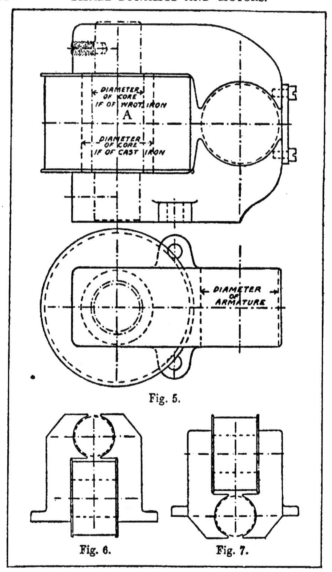

Fig. 5.

Fig. 6.   Fig. 7.

done, and the core may then be made from about two-thirds to four-fifths of the diameter of the cast core, as given. These relative diameters are shown in fig. 5. The diameter of the cast iron core is shown with a plain "dash" line; that of the wrought iron core with a "dot and dash" line. The diameters of armatures in this class should be equal to their lengths. If an armature is required to be longer than its diameter, the width of field magnet is increased, and the cross-sectional areas of core should also be increased in a like proportion.

The second group (figs. 8, 9 and 10) represents the various forms taken by a very popular class of dynamo magnets. It is wise to have each of these also cast in a single piece, although they are somewhat difficult to wind when that plan is adopted. Joints, if made, should be at the place marked B. This type of carcase cannot very well be made partly of cast and partly of wrought iron—which would involve too many joints. A popular form is, however, embodied in fig 11, which shows one of the "Kapp-type" machines in three forgings. This requires a slightly different winding from the other machines in this class, and particulars are given specially on page 46. The armatures for machines of the second class may be any reasonable length, since, as the field magnet cores are flat, they can be made any width without altering their proportional sectional areas. They are, however, very well suited to drum-type armatures about half as long again as

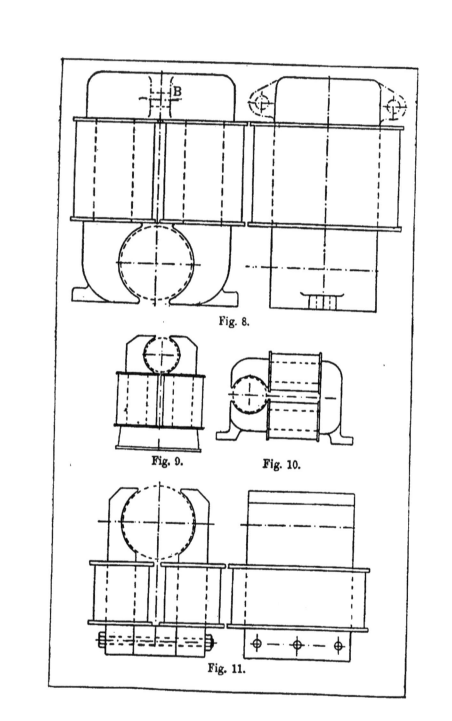

Fig. 8.

Fig. 9.

Fig. 10.

Fig. 11.

their diameters, and these proportions form the basis of the calculations for Chapter V. If the length of an armature is made less than this proportion, the field magnet cores can very well be made circular in cross-section, thus approximating in appearance to the Edison-Hopkinson dynamo. The cross-sectional area must, of course, be made equivalent to that given for a machine of equal size with flat cores. Shuttle, tri-polar, or quadri-polar armatures, if used at all, should be employed in field magnets of this class.

"Manchester"-type dynamos, two forms of which are shown in figs. 12 and 13, are also two-coil machines. They are generally supposed to be fitted with Gramme-ring armatures, as they are best suited to accommodate armatures of large diameter as compared with their length. Fig. 13 shows the proportions of a machine designed with that object, the length of the ring being two-thirds the diameter. Fig. 12 shows a Manchester dynamo taking a drum armature of same diameter as length, and for the smaller machines drum armatures are recommended to be used in preference to Gramme rings. Manchester-type dynamos almost invariably have cast iron yokes (C, figs. 12 and 13), and round wrought iron cores; the proportions here given are to suit such construction. A very efficient machine could be constructed as in fig. 14, being all of cast iron, with joints at D. It may be remarked that as these constitute the ends of the magnet poles, joints here will not affect the magnetic flux, whilst the two parts will be easy to wind, and the joints do not need to be made with special accuracy.

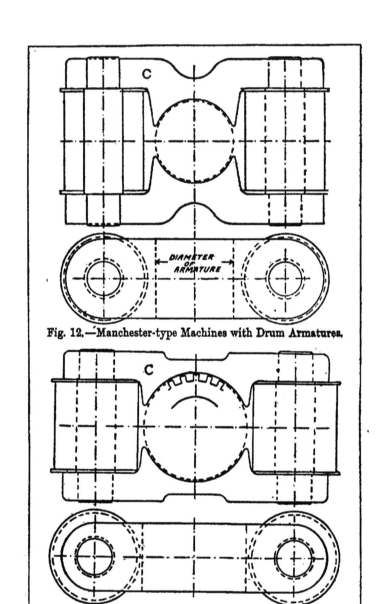

Fig. 12.—Manchester-type Machines with Drum Armatures.

Fig. 13.—Manchester-type Machines with Ring Armatures.

The four-coil magnet (fig. 15) is included, because it is a useful type for good-sized motors, and if each half is centered in the lathe, it will be found particularly easy to wind the coils. Windings are given for machines of this type on page 44, and drum armatures

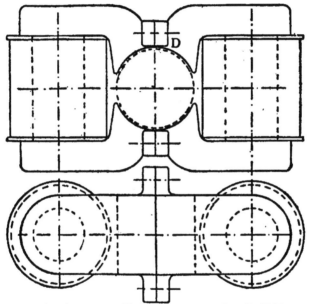

Fig. 14.—Manchester-type Machines with Cast Iron Field Magnets.

having length equal to diameter are understood to be employed.

Belonging to each group of drawings in this chapter is a set of scales. These are for the purpose of setting out to full size the different machines for any particular output. A machine can be designed straight away from these drawings for either 10, 20, 30, 40, 60, 100, 150, 250, 400 or 500 watts output, the field magnets being drawn correctly by means of the proper scale, and suitable windings being found in the tables in Chapter V. The first set of scales on page 21 refers

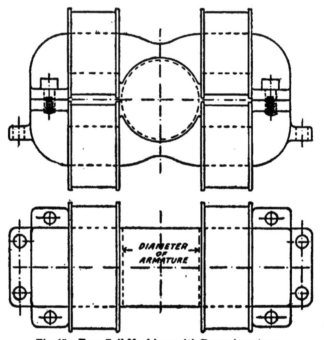

Fig. 15.—Four-Coil Machines with Drum Armatures.

to figs. 5, 6, 7, 12, 14 and 15; the next to figs. 8, 9, 10 and 11; and the last to fig. 13 only. The reader should take care not to be confused when using one of the scales, or mistakes will easily arise. If he desires to make a drawing full size of one of the field

Scales for Machines as in Figs. 5, 6, 7, 12, 14 and 15.

Scales for Machines as in Figs. 8, 9, 10 and 11.

Scales for Manchester-type Machines with Ring Armatures.
(Fig. 13.)

magnets for a machine of say 40-watts output he should first transfer the "40-watt scale" to the edge of a strip of stout paper or Bristol board, and use this to scale off all dimensions from the drawing. Of course hte proper scale must be used, as indicated by the references marked on each drawing. The smaller diagrams on pages 14 and 16 are to half the scale of the representative desgins, but should be regarded as diagrammatic only. They show the alternative methods of fitting up each type of machine, the essential points of cross sectional area, etc., being retained in each form.

CHAPTER III.

ARMATURES: DIFFERENT TYPES AND HOW TO WIND THEM.

THE simplest type of armature is that shown in fig. 16 and known as the "Siemen's H." Beyond its simplicity it has nothing to recommend it, but as that is a quality of prime importance to the beginner it cannot

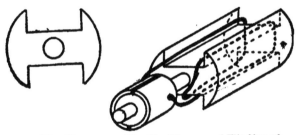

Fig. 16. Fig. 17.—Diagram of Winding of H Armature.

very well be neglected. The method of winding is shown in fig. 17, one end of the wire being connected to each half of the commutator. Care must be taken when winding a Siemen's armature not to reverse the winding, which it is possible to do when turning the

23

armature round. Such armatures should only be used
in the smallest machines, or in "first attempts." Siemen's armatures cannot be used for dynamos required
to charge accumulators, and motors fitted with this
type are not self-starting.

Tri-polar and quadri-polar armatures are of use for
small self-starting motors. These can be wound in
two ways ("open" and "closed" circuit), both being
equally effective, and the latter is illustrated in fig. 18.
In winding a tri-polar armature in this way, each

Fig. 18.—Diagram of Winding of a Tripolar Armature.

limb is wound in exactly the same direction, the beginning and end of each coil being carefully marked.
Taking the three poles in rotation, the beginning of
coil 1 and the end of coil 3 are twisted together,
"tinned," and both then soldered to one of the commutator segments (*a*). The end of coil 1 and beginning
of coil 2 are then similarly treated, and joined to commutator bar (*b*). End of coil 2 and beginning of No.
3 are joined to the remaining segment (*c*). The three
coils now practically become a single unbroken wire,

with three connections leading to the three commutator segments.

The winding for a "quadri-polar" armature is precisely similar in principle to that of the tri-polar armature last described. A "quadri-polar" armature is not recommended. It is quite as easy to wind a drum armature of 8 slots in 4 sections (shown in fig. 42), or a simple 4-part drum armature as shown in fig. 19.

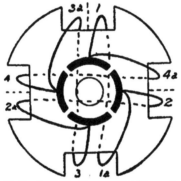

Fig. 19.—Method of Winding a 4-Slot Drum Armature.

The above-mentioned armatures are sometimes of cast iron. These of course are not admissible in any but the beginner's first attempt, laminations or stampings being in every way superior. All the armatures next to be described are built of stampings, which are obtainable in practically any size and with any number of slots.

All armatures for practical machines, even the smallest, should be of the drum or Gramme-ring type. The latter, generally speaking, are the simplest to wind, but for small machines, drum armatures are best. A ring armature of correct proportions under 3 inches diameter is difficult to wind: consequently it is not recommended that machines of less capacity than 60 watts be fitted with Gramme-ring armatures.

Drum armatures should have the number of teeth somewhat proportioned to the size. As a rough guide, the diagrams on page 29, showing various sizes of armatures for the machines described in this book, will be useful. Any drum armature with an even number of slots can be wound either in the same number of sections as there are slots or in half the number. The greater number is better, but more difficult; and the best plan is therefore to compromise in the larger sizes, by crowding in a larger number of slots and winding half the number of sections. Diagrams for winding in various approved ways and for armatures having different numbers of slots are given, so that the reader may have a choice of examples (figs. 41 to 49). Generally speaking, when winding two sections in one pair of slots, the easiest plan is to fit a slip of wood just half the width of winding space in each slot and to fill up the remaining space with wire. But the sections are often wound one on top of the other, and there is little difference as regards efficiency. As far as possible make every section equal as regards amount and number of turns of wire in it.

The length of an armature in proportion to its diameter is of some importance. In the series of dynamos dealt with in this book these proportions have been laid down (see Chapter II), and any considerable departure from them involves an alteration in the field-magnet designs and windings; but slight variations can be made by the intelligent amateur, for whose benefit it may be remarked that none of the examples laid down in this book are absolutely binding. Adherence to them will enable those not deeply acquainted with the subject to turn out well proportioned and correctly wound machines, but as long as the *ratio* of field-magnet winding to armature winding is kept as given herein, a slightly longer or shorter armature may be used to obtain a corresponding slight alteration of output. The field-magnet casting itself should also be altered to suit the new armature, being made wider or narrower, as the case may be.

Figs. 20 to 35 show cogged drum armature stampings from 1⅛ to 4¼ inches diameter. Each figure shows three of the slots (full size), the number of slots being placed in the middle of the three, the width of slot in the left and its depth in the right hand of the other slots shown. While it is recommended that these dimensions and numbers of slots be adhered to for the sizes quoted, any equivalent "slot-area" will accommodate the same amount of wire, the tables in Chapter V having been prepared on the basis of the diagrams here given. When ordering stampings, if any doubt exists as to their thickness, the price "per inch in depth" should be asked, and it should be remembered that paper insulation between the stampings will increase the depth to a slight extent. In smaller sizes, paper is altogether unnecessary, but the stampings may all be

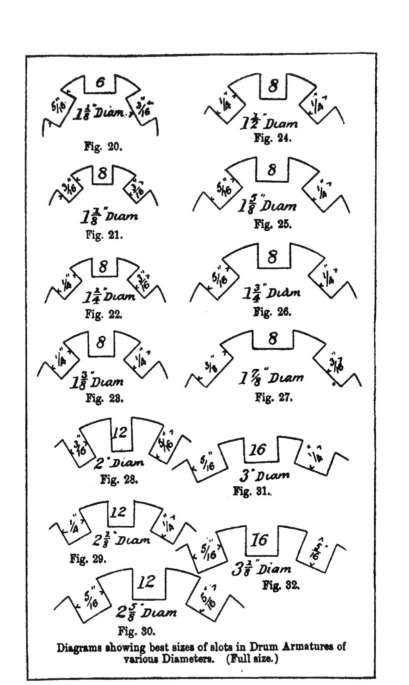

Fig. 20.

Fig. 21.

Fig. 22.

Fig. 23.

Fig. 24.

Fig. 25.

Fig. 26.

Fig. 27.

Fig. 28.

Fig. 29.

Fig. 30.

Fig. 31.

Fig. 32.

Diagrams showing best sizes of slots in Drum Armatures of various Diameters.   (Full size.)

spread out on a board and a thin coat of any ordinary varnish, paint, or enamel applied on one side only. This

Fig. 33.

is absolutely effective for the prevention of eddy currents, and is a very simple method to adopt.

Fig. 34.　　　　　Fig. 35.

Ring armatures above 3 inches diameter may be plain stampings if preferred, with notches for fibre

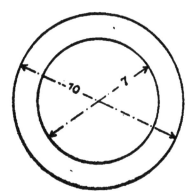

Fig. 36.—The Best Proportions for a Gramme Ring.

driving horns, or special brass end spiders with teeth may be employed. The ratio of the external to internal diameter is best about 10 to 7, but in the smaller sizes 10 to 8 is a reasonable ratio, and makes the armature easier to wind. The best proportion (10:7) is indicated i nthe diagram, fig. 36. The proper winding of a ring armature is very easily grasped, and the diagram (fig. 37) should make the principle perfectly clear to those unacquainted with it.

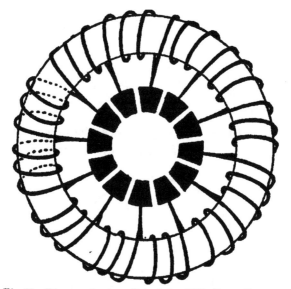

Fig. 37.—Diagram to show Principle of Winding a Gramme Ring Armature.

With plain ring armatures the winding should never be more than three layers deep. The tables in Chapter V (page 43) provide for this, and the point should never be departed from. Figs. 38, 39 and 40 show the recommended proportions of cogged Gramme rings for 60, 100 and 150 watts respectively, and are designed

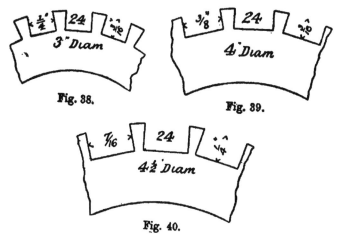

Fig. 38.

Fig. 39.

Fig. 40.

Proportions of Ring Armatures for Manchester Type Machines,
Fig. 13, for 60, 100, and 150-watts output respectively.

to suit the tables of wiring on page 51. For higher outputs, plain rings are advisable.

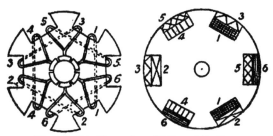

Each Coil is Shown by a Different Section.

Fig. 41.—Diagram of winding of a 6-slot Armature.

There is little difficulty in following the winding dia-

gram of a drum armature if it is first studied carefully.
Indeed, it is generally easier to understand a proper

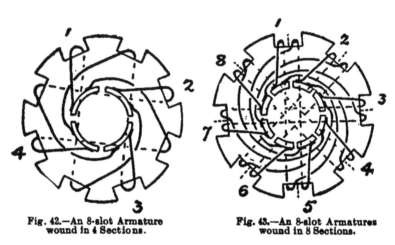

Fig. 42.—An 8-slot Armature
wound in 4 Sections.

Fig. 43.—An 8-slot Armatures
wound in 8 Sections.

diagram than a written description, and the drawings
(figs. 41 to 49), which show methods of winding drum

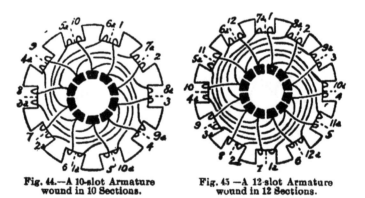

Fig. 44.—A 10-slot Armature
wound in 10 Sections.

Fig. 45 —A 12-slot Armature
wound in 12 Sections.

armatures in different ways and with differing numbers
of slots, will be understood if a study of them is made

in the order given. "When in doubt" try winding a "dummy" armature, consisting of a cotton reel or

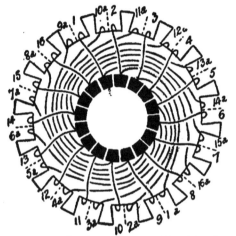

Fig. 46—A 16-slot Armature wound in 16 Sections.

bobbin with "slots" cut in the ends, string being good

Fig. 47.—A 24-slot Armature wound in 12 Sections.

enough for the experiment. Mark the beginnings and

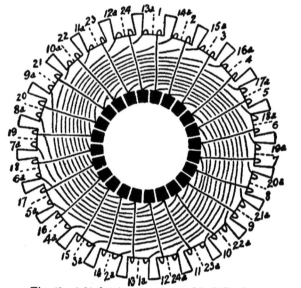

Fig. 48.—A 24-slot Armature wound in 24 Sections.

Fig. 49.—Diagram of winding of Armature for a 4-pole Machine.

endings and the order of the wires, and a mistake will not easily be made.

The method of winding a drum armature for a 4-pole machine with alternate N and S poles is shown in fig. 49. It will be noted that the winding is not in opposite slots, but in those at 90° apart. A large number of sections is essential, and the commutator should have the opposite segments joined together (cross-connected) as shown, when two brushes only need be employed, being set—not diametrically opposite, as is usual in bi-polar machines—but at right angles to one another.

CHAPTER IV.

COMMUTATORS AND OTHER DETAILS.

THE commutators for small dynamos, except for Siemen's and tri-polar armatures, differ principally from those of larger machines in the material of which the bars are composed. Copper is the metal invariably chosen for large dynamos, while brass, owing to its superiority for turning purposes, is generally used for small generators and motors. The method of construction is different only in the smallest sizes, although as a matter of fact the principle on which a large commutator is built up is quite suitable for copying in a model of fair size.

In fig. 50 is shown a 2-part commutator such as would be fitted to a "Siemen's H" armature. It consists essentially of a ring or tube of brass, mounted on a wooden or "fibre" core, split lengthwise on two opposite sides, and each part insulated from the other. The acknowledged method of construction of such a commutator is as follows: A short length of hard wood or vulcanized fibre is drilled with a hole to fit tight in its correct position on shaft, and when so fixed, the shaft or spindle is placed in the lathe. The

wood core is then turned down until a suitable piece of
brass tube, which should be of correct circular section,
can be driven on easily.   The brass tube having been
mounted in this way, is marked (by means of a scriber
along the top of the lathe T-rest) with two lines diametrically opposite, as shown by the two dots in the
end view of the commutator (fig. 52).   A short distance away from each of these lines and on each side
of them are marked four other longitudinal lines.   The
centre line and the line on each side of it are shown
in fig. 52 at *a, b* and *c.*   The lathe can then be turned

Fig. 50.   Fig. 51.

Commutators for Siemen's H Armatures.

slowly, and two more lines marked as at *d* and *e,* cir-
cumferentially; then where *d* and *e* cross *b* and *c*
(and also in the corresponding places on the other
side of the tube) holes are drilled and countersunk for
very small ordinary wood screws.

It is important to note that on no account must either
of the eight screws reach down to the spindle running
through the commutator.   The result would generally
mean a short circuit.   Screws must, therefore, be
chosen of such a length that even when driven home
they clear the shaft by a reasonable distance; thus with

a shaft $\frac{3}{16}$ inch diameter, and the cummutator tube $\frac{3}{4}$ inch diameter outside, $\frac{1}{4}$ inch screws would just clear by $\frac{1}{32}$ inch. This would be running matters rather close, and to make sure, the screws should have the extreme points filed off.

Next to the importance of the above is the necessity of having the commutator "true." Not only must the tube forming the segments be externally at all events truly circular, but it must be fitted absolutely central; the screw heads must not project, and the surface should be quite smooth. When this has been accomplished, the screws being driven properly home, the

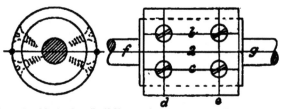

Fig. 52.—Method of Building a simple two-part Commutator.

tube must be split by means of a metal saw or very fine "key" file along the line f g, on both sides, as shown at the two dots on the left hand view of fig. 52. The saw cut must be continued right through the brass and slightly into the insulating core, but not so much as to weaken this latter. The commutator now consists of two separate and insulated segments of equal size, and the spaces between them may—and should—be filled with well-fitting slips of wood, fibre, or even cardboard glued in.

Sometimes the saw cuts in a two-segment commutator are cut diagonally, as indicated in fig. 51. When this is done, the brushes bear on both segments at once during the brief period of commutation, and this theoretically is advantageous, because the sudden commutation which takes place with the "straight-across" variety is apt to augment sparking. If the slots are cut narrow, however, there is not much—if any—to choose betweeen the two commutators. Possibly fig. 51 looks a little more "professional," for those who study appearances.

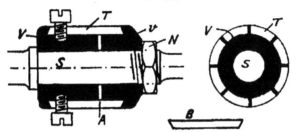

Fig. 53—Commutator for a Small Machine.

The commutator is perhaps the most difficult part of a dynamo for the amateur to manipulate. It is most decidedly not an easy matter to build a really good sound commutator, even in the smallest size, without the help of a lathe. The advice, so often freely given, to "mount a round piece of wood on the spindle, drive on a short length of brass tube," etc., serves only to reveal ignorance of the practical subject, and certainly must have proved exasperating to many hundreds of beginners—who quickly realize the tendency of the round piece of wood to wobble, to run eccentrically, to slide freely about the spindle,

and do everything a commutator should not. Yet
the writer has seen a dynamo whose commutator
ran true and firm though made from a portion of a
cotton reel, mounted on a bush of brass tube, and this
driven on a piece of silver-steel rod. It should be
said that the maker of this little machine was not
a "beginner."

Drawings of a correctly-made commutator for a
small dynamo with armature wound in eight sections
are here given (fig. 53). It will be seen at once that
the chief difference between this and the previous
examples is the method by which the segments are
secured. No screws pass through them for that
purpose at all. On the spindle S, against a shoulder
turned at the correct place, is tightly driven a piece
of vulcanized fibre or boxwood V. A second piece of
the same material, v, is similarly driven on, and both
are secured by the nut running on the screwed part
of spindle, N. The core is then turned with the
lathe running at a high speed, the central portion being
cut away until its diameter is correct for the internal
diameter of a suitable piece of thick brass tubing, or
solid rod drilled out, T. The two ends are larger,
being almost equal to the outside diameter of the
tubing, and the shoulder at each inner end is undercut
at a sharp angle, as shown.

All this having been done, the brass tubing may be
mounted on a wooden mandrel and have its edges
turned down as nearly as possible to the angle of the
shoulders in the core. The length of the tube should
be slightly greater than the distance between the

shoulders, so that when inserted a small space will be left, as at A. The tube may then be marked out into eight spaces, the pieces cut apart with the saw, burrs removed with the least possible amount of filing, . and the segments fitted at equal distances around the core, the nut and piece v being run back to allow this to be done. Eight strips of fibre or wood of the form shown at B should be prepared, the proper thickness to fill the saw cuts, and when these are in place, and

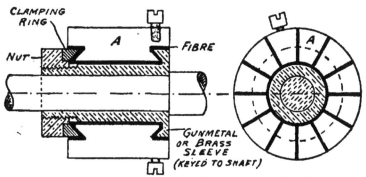

Fig. 64.—Commutator of Ordinary Type for Larger Machines.

the nut tightened up, the commutator should be rigid and should run quite true.

In larger dynamos the practice is either to cast the segments—or bars—of copper, or to make them from rolled sections of that metal. The shapes of these bars are often very peculiar, but it is not here intended to go into the construction of machines requiring any special forms of commutator. Diagrams are given in figs. 54 and 55 to show the construction of a 12-part

commutator, and a description is unnecessary, since it
is self-evident that the principle in this is the same
as in the last case, the difference being the form of the
bars and the addition of the lugs for armature connec-

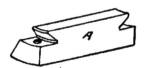

Fig. 55.—Form of Bars in Commutator of Type shown in Fig. 54.

tions. The bars should be cast in copper or brass, or
filed to shape out of suitable rods. The latter is a
tedious operation, easily resulting in failure, and therefore taxing the care and skill of the maker.

CHAPTER V.

TABLES OF WINDINGS.

A SEPARATE table of windings is not given for the special machine illustrated in fig. 14, page 19. The armature proportions for such a machine will be precisely as for an ordinary Manchester-type (fig. 12) of similar output, and the windings may be taken as exatcly the same as those for C-type machines (figs. 5, 6 and 7), given on page 44.

Windings are only given for ring-wound Manchester dynamos and motors over 60 watts. All these require special armatures, which are illustrated in figs. 38 to 40. The proper fig. number will be found in the "output" columns, page 49.

WINDINGS FOR MACHINES SHOWN IN FIGS. 5, 6 AND 7. PAGE 14. ARMATURES ; laminated cogged drums, as figs. 20 to 35. Diameter equal to length. FIELD-MAGNET COILS IN SHUNT.

Output.	Volts.	Ampères	Wire for Armature.	Wire for Field Magnets.
10 watts	5 10	2 1	1½ oz. No. 22 1½ oz. No. 24, 25 yd	8 oz. No. 22 shunt 8 oz. No. 24 shunt
20 watts	5 10 20	4 2 1	2¾ oz. No. 20 2½ oz. No. 22 2 oz. No. 26	1 lb. No. 19 1¼ oz. No. 22 1 lb. No. 25
30 watts	5 10 20	6 3 1·5	4⅛ oz. No. 18 4½ oz. No. 20 5¾ oz. No. 22	1½ lb. No. 18 1 lb. 14 oz. No. 20 2 lb. No. 23
40 watts	5 10 20	8 4 2	5¾ oz. No. 18 5¼ oz. No. 20 5½ oz. No. 23	2¾ lb. No. 17 3 lb. No. 19 2¾ lb. No. 22
60 watts	10 15 30	6 4 2	6–7 oz. No. 19 7¼ oz. No. 20 6¾ oz. No. 23	2¾ lb. No. 18 2½ lb. No. 20 3 lb. No. 22
100 watts	20 50 100	5 2 1	14 oz. No. 19 14 oz. No. 22 9½ oz. No. 26	3 lb. 10 oz. No. 20 2¾ lb. No. 24 2 lb. 6 oz. No. 28
150 watts	30 50 100	5 3 1·5	16 oz. No. 20 15½ oz. No. 23 14 oz. No. 24	6 lb. No. 20 6¼ lb. No. 22 7 lb. No. 24
250 watts	30 50 100	8·3 5 2·5	1½ lb. No. 18 1¼ lb. No. 20 1 lb. 6 oz. No. 22	9½ lb. No. 19 12 lb No. 20 12 lb. No. 23
400 watts	50 60 100	8 6.6 4	31 oz. No. 18 1¾ lb. No. 19 1 lb. 10 oz. No. 21	12 lb. No. 20 12 lb. No. 21 12 lb. No. 22
500 watts	50 60 100	10 8·3 3	2¼ lb. No. 17 2 lb. No. 18 1 lb. 14 oz. No. 20	15 lb. No. 19 15 lb. No. 20 15 lb. No. 22

WINDINGS FOR MACHINES SHOWN IN FIGS. 8, 9 AND 10. PAGE 16, ARMATURES: laminated cogged drums, as figs. 20 to 35. Diameter two-thirds of length. FIELD-MAGNET COILS IN SHUNT.

Output.	Volts.	Ampères	Wire for Armature.	Wire for Field Magnets.
10 watts	5 10	2 1	1¾ oz. No. 22 1½ oz. No. 26	6 oz. No. 22 5½ oz. No. 26
20 watts	5 10 20	4 2 1	2⅛ oz. No. 20 2⅓ oz. No. 22 2 oz. No. 26	10 oz. No. 20 14 oz. No. 22 10 oz. No. 26
30 watts	5 10 20	6 3 1·5	5 oz. No. 18 4¼ oz. No. 20 4 oz. No. 23	1⅛ lb. No. 18 1¼ lb. No. 21 1¾ lb. No. 23
40 watts	5 10 20	8 4 2	6½ oz. No. 17 4½ oz. No. 20 4 oz. No. 22	1⅓ lb. No. 18 1¾ lb. No. 20 2½ lb. No. 22
60 watts	10 15 30	6 4 2	6½ oz. No. 18 5 oz. No. 20 5¾ oz. No. 22	2¼ lb. No. 19 2⅝ lb. No. 20 2¾ lb. No. 22
100 watts	20 30 50	5 3·3 2	8 oz. No. 20 9 oz. No. 21 8½ oz. No. 23	4 lb. No. 20 3¾ lb. No. 21 3⅛ lb. No. 23
150 watts	30 50 100	5 3 1·5	11 oz. No. 20 13 oz. No. 21 12½ oz. No. 24	6 lb. No. 20 6 lb. No. 22 6½ lb. No. 25
250 watts	30 50 100	8·3 5 2·5	1⅓ lb. No 17. 14 oz. No. 20 17 oz. No. 22	9 lb. No. 19 8½ lb. No. 21 7½ lb. No. 24
400–500 watts	50 60 100	10 8·3 5	2½ lb. No. 17 2½ lb. No. 18 2 lb. No. 20	12½ lb No. 20 11 lb. No. 21 11 lb. No. 25

WINDINGS FOR MACHINES SHOWN IN FIGS. 8, 9, 10 (page 16), but with H (Siemen's) Armatures. FIELD-MAGNET COILS IN SHUNT.

Output.	Volts.	Ampères	Wire for Armature.	Wire for Field Magnets.
10 watts	5 10	2 1	1¾ oz. No. 20 2 oz. No. 22	6 oz. No. 22 5½ oz. No. 26
20 watts	5 10 20	4 2 1	4 oz. No. 20 3 oz. No. 22 3 oz. No. 24	10 oz. No. 20 14 oz. No. 22 10 oz. No. 26
30 watts	5 10 20	6 3 1·5	4 oz. No. 18 4 oz. No. 20 5 oz. No. 22	1½ lb. No. 18 1¼ lb. No. 21 1¾ lb. No. 23
30 watts	5 10 20	8 4 2	5 oz. No. 17 5 oz. No. 19 5½ oz. No. 21	1⅛ lb. No. 18 1¾ lb No. 20 2½ lb. No. 22

WINDINGS FOR MACHINES SHOWN IN FIGS. 8, 9 AND 10, with Siemen's H Armatures, and FIELD-MAGNET COILS IN SERIES.

Output.	Volts.	Ampères	Wire for Armature	Wire for Field Magnets.
10 watts	5	2	1¾ oz. No. 20	5 oz. No. 18
	10	1	2 oz. No. 22	6 oz. No. 19
20 watts	5	4	4 oz. No. 20	8 oz. No. 18
	10	2	3 oz. No. 22	10 oz. No. 20
	20	1	3 oz. No. 24	9 oz. No. 21
30 watts	5	6	4 oz. No. 18	10 oz. No. 16
	10	3	4 oz. No. 20	10 oz. No. 18
	20	1·5	5 oz. No. 22	10 oz. No. 20
40 watts	5	8	5 oz. No. 17	1 lb. No. 14
	10	4	5 oz. No. 19	1¼ lb. No. 16
	20	2	5½ oz. No. 21	1 lb. 2 oz. No. 18

WINDINGS FOR "KAPP"-TYPE MACHINES (AS IN FIG. 11). With soft wrought iron field magnets. ARMATURES: laminated cogged drums. Diameter two-thirds of length. FIELD-MAGNET COILS IN SHUNT.				
Output.	Volts.	Ampères	Wire for Armature.	Wire for Field Magnets.
10-watts	5 10	2 1	1¾ oz. No. 22 1½ oz. No. 26	4½ oz. No. 24 5 oz. No. 27
20-watts	5 10 20	4 2 1	2½ oz. No. 20 2⅛ oz. No. 22 2 oz. No. 26	7 oz. No. 22 10 oz. No. 24 8 oz. No. 27
30-watts	5 10 20	6 3 1·5	5 oz. No. 18 4¼ oz. No 20 4 oz. No. 23	1 lb. No. 20 1 lb. No. 22 1¼ lb. No. 24
40-watts	5 10 20 .	8 4 2	6½ oz. No. 17 4½ oz. No. 20 4 oz. No. 22	1 lb. No. 19 1⅛ lb. No. 22 1⅜ lb. No. 24
60-watts	10 15 30	6 4 2	6½ oz. No. 18 5 oz. No. 20 5¾ oz. No. 22	1¾ lb. No. 21 2 lb. No. 21 2 lb. No. 24
100-watts	20 30 50	5 3·3 2	8 oz. No. 20 9 oz. No. 21 8½ oz. No. 23	3⅛ lb. No. 21 3 lb. No. 22 3 lb. No. 24
150-watts	30 50 100	5 3 1·5	11 oz. No. 20 13 oz. No. 21 12½ oz. No. 24	4¾ lb. No. 21 4 lb. No. 24 4½ lb. No. 26
250-watts	30 50 100	8·3 5 2·5	1⅛ lb. No. 17 14 oz. No. 20 17 oz. No. 22	7½ lb. No. 30 7 lb. No. 22 6½ lb. No. 25
4 00-500 watts	50 60 100	10 8·3 5	2¼ lb. No. 17 2½ lb. No. 18 2 lb. No. 20	10 lb. No. 21 8½ lb. No. 22 7 lb. No. 25

WINDINGS FOR MACHINES SHOWN IN FIGS. 12 AND 15, PAGES 18 and 21. ARMATURES : laminated cogged drums, diameter equal to length. FIELD-MAGNET COILS IN SHUNT.

Output.	Volts.	Ampères	Wire for Armature.	Wire for Field Magnets.
10 watts	5	2	1½ oz. No. 22	7½ oz. No. 22
	10	1	1½ oz. No. 24	7¼ oz. No. 24
20 watts	5	4	2¾ oz. No. 20	15 oz. No. 19
	10	2	2½ oz. No. 22	13 oz. No. 22
	20	1	2 oz. No. 26	15 oz. No. 25
30 watts	5	6	4½ oz. No. 18	23 oz. No 18
	10	3	4½ oz. No. 20	29 oz. No. 20
	20	1·5	5¾ oz. No. 22	1 lb. 15 oz. No. 23
40 watts	5	8	5¾ oz. No. 18	2 lb. 10 oz. No. 18
	10	4	5½ oz. No. 20	2 lb. 14 oz. No. 20
	20	2	5¼ oz. No. 23	2 lb. 10 oz. No. 23
60 watts	10	6	6–7 oz. No 19	2 lb. 10 oz. No. 19
	15	4	7¼ oz. No. 20	2 lb. 6 oz. No. 21
	30	2	6¾ oz. No. 23	2 lb. 14 oz. No. 23
100 watts	20	5	14 oz. No. 19	3 lb. 7 oz. No. 21
	50	2	14 oz. No. 22	2 lb. 10 oz. No. 25
	100	1	9½ oz. No. 26	2 lb. 4 oz. No. 29
150 watts	30	5	16 oz. No. 20	5¾ lb. No. 21
	50	3	15½ oz. No. 23	6 lb. No. 23
	100	1·5	14 oz. No. 24	6½ lb. No. 25
250 watts	30	8·3	1½ lb. No. 18	9 lb. No. 20
	50	5	1¼ lb. No. 20	11½ lb. No. 21
	100	2·5	1 lb. 6 oz. No. 22	11½ lb. No. 24
400 watts	50	8	31 oz. No. 18	11¾ lb. No. 22
	60	6·6	1¾ lb. No 19	11½ lb. No. 22
	100	4	1 lb. 10 oz. No. 21	11¼ lb. No. 21
500 watts	50	10	2¼ lb. No. 17	14½ lb. No. 21
	60	8·5	2 lb. No. 18	15 lb. No. 21
	100	5	1 lb. 14 oz. No. 20	14½ lb. No. 23

WINDINGS FOR MANCHESTER TYPE MACHINES, SHOWN IN FIG. 13. RING ARMATURES, diameter being half as great again as length. FIELD-MAGNET COILS IN SHUNT.

	Output	Volts.	Ampères	Wire for Armature.	Wire for Field Magnets.
See Fig. 40|See Fig. 39|See Fig. 38	60 watts	10 15 30	6 4 2	11 oz. No. 18 9 oz. No. 20 9¼ oz. No. 23	2¾ lb. No. 18 2½ lb. No. 20 3 lb. No. 22
	100 watts	20 50 100	5 2 1	15 oz. No. 19 17 oz. No. 22 1¼ lb. No. 24	3 lb. 10 oz. No. 20 2¾ lb. No. 24 2 lb. 6 oz. No. 28
	150 watts	30 50 100	5 3 1·5	1¾ lb. No. 18 1¾ lb. No. 20 1¼ lb. No. 24	6 lb. No. 20 6¼ lb. No. 22 7 lb. No. 24
	250 watts*	30 50 100	8·3 5 2·5	1¾ lb. No. 18 (2 layers) 1½ lb. No. 20 (2 layers) 1¾ lb. No. 22 (3 layers)	9½ lb. No. 19 12 lb. No. 20 12 lb. No. 23
	400 watts†	50 60 100	8 6·6 4	2⅛ lb. No. 18 (2 layers) 1¾ lb. No. 20 (2 layers) 2⅛ lb. No. 21 (3 layers)	12 lb. No. 20 12 lb. No. 21 12 lb. No. 25
	500 watts‡	50 60 100	10 8·3 5	2½ lb. No. 18 (2 layers) 2½ lb. No. 19 (2 layers) 2¼ lb. No. 21 (3 layers)	15 lb. No. 19 15 lb. No. 20 15 lb. No. 22

* Armature a plain ring, 5 ins. diameter outside and 3½ ins. inside, x 3 ins. wide. To be wound in 24 sections. Armature tunnel to be bored out to suit—5⅝ ins. diameter should be right.

† Armature a plain ring, 5¼ ins. diameter outside and 3⅝ ins. inside, x 3¼ ins. wide. To be wound in 24 sections. Armature tunnel to be bored to about 5⅝ ins. diameter.

‡ Armature a plain ring, 5½ ins. diameter outside, 3⅝ ins. inside, x 3½ ins. wide. To be wound in 24 sections. Armature tunnel bored out to 5⅝ ins. diameter.

CHAPTER VI.

How to Build a Small Machine.

A few notes on the construction of a small machine will be of use to those who may be styled beginners, and although a full description of the proper method of building a large dynamo hardly comes within the scope of this little book, it may be remarked that all the directions as regards insulation, etc., apply as much in the one case as in the other.

The reader is supposed to have decided on building a machine of the "Simplex," or single-coil type, with armature beneath, as in fig. 7. He decides that it shall be of 10-watt capacity, say, 2 amperes at 5 volts, for charging a small 4-volt pocket accumulator.

The first operation is to make a full size drawing of the field-magnet, showing an end elevation and a longitudinal sectional elevation, as shown full size in figs. 56 and 57, and for this purpose the "10-watt" scale at top of page 23 must be used, when it will at once be seen that an armature 1¼ ins. diameter × 1¼ ins. long will be required. On referring to the chapter on

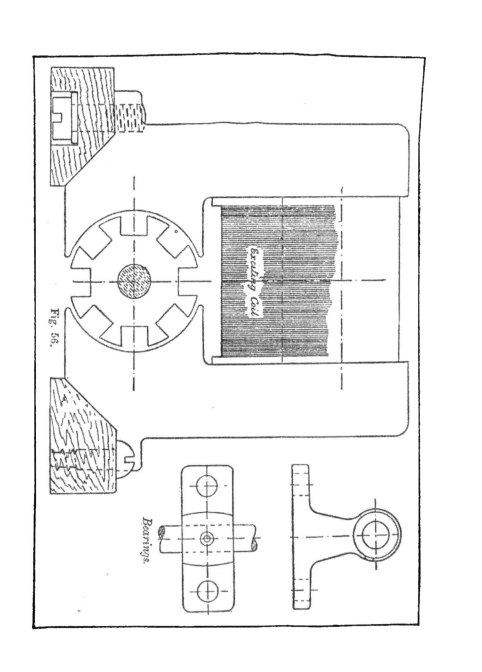

Fig. 56.

Exciting Coil

Bearings.

armatures it will be seen that the armature, if of the drum type, should have eight slots ¼ in. × 3⁄16 in. A drum-type armature is recommended in preference to a tri-polar one, which is the only other permissible, since the machine is to charge storage cells. The drawing is completed as shown in figs. 56 and 57.

The spindle is turned from a piece of 3⁄8-in. silver steel rod. A shoulder of the full 3⁄8-in. diameter is left for the stampings to back against, and a portion of the 5⁄16 in. part is screwed with a fine thread.

The armature tunnel is bored (if it has not been accurately cored out) to 1⁄16 in. larger in diameter than the finished outside diameter of the armature core. Great care is necessary when boring out the field magnets not to break them apart by taking too heavy a cut. It may be necessary to secure the two poles by screwing to both of them a plate of brass or iron temporarily. In fig. 5 such a plate is shown, joining and strengthening the poles, but it should be remembered that if the plate is to be a permanent fixture it *must not be iron.* A brass plate in the position shown in fig. 5 is a protection to the armature of a machine of this type, and may be engraved with the name of the builder, etc.

Presuming the stampings are circular and true (as they should be), little will have to be done to them in the way of "truing up" after they are assembled. When this is done, and the commutator fitted, the important item of insulation must be attended to. For all machines, the best materials are red vulcanised fibre (or "fibre," as it is commonly termed) and mica. For very small armatures, however, neither of these mater-

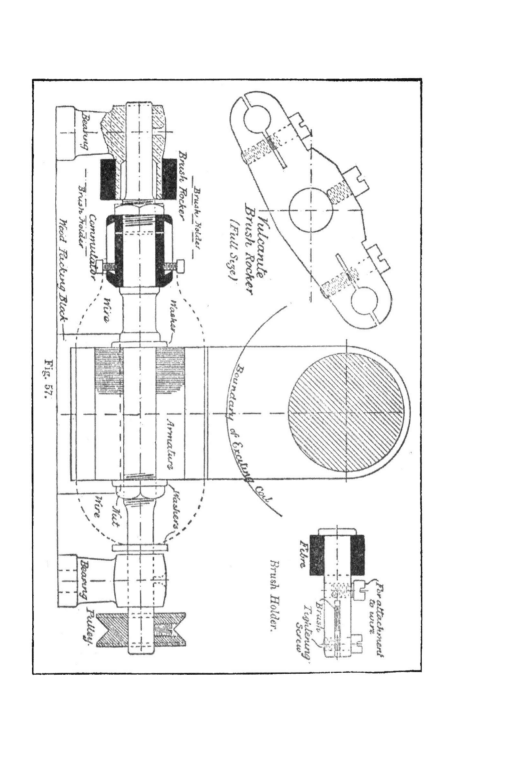

Vulcanite
Brush Rocker
(Full Size)

Boundary of Exciting Coil

Brush Holder.

Fibre

Brush
Tightening
Screw

For attachment
to wire

Brush Holder

Brush Rocker

Commutator

Brush Holder

Wood Rocking Block

Wire

Washers

Armature

Washers

Wire

Nut

Bearing

Bearing Pulley.

Fig. 57.

ials is easy to fix, and the thinnest obtainable is really hardly thin enongh. If fibre is used, it may be soaked in water, when it will take sharp bends without cracking; it is fixed with good glue or very thick shellac varnish.

A better plan for an armature of the size under consideration is to insulate with brown paper *soaked* with paste till quite soft. Every part of the armature and the shaft up to the commutator one way and for an inch the other way must be covered with insulation. A very good alternative to brown paper is narrow "sarcenet" ribbon, obtainable at any milliner's; pieces being cut to right lengths and glued on. The armature must be left to dry before winding—the insulation is almost certain to be abraded if time is not given for this.

The armature should next be wound in any approved way—the easiest in the present instance being in four sections for a 4-part commutator. Reference to the table on page 46 will show that for this machine will be required 1½ ozs. of No. 22 wire for the armature, and 8 ozs. of No. 22 for the field magnets, which must be connected in shunt. After winding the armature, careful tests must be made to ensure the wire being thoroughly insulated from the iron core. If all is well, paint the whole armature with thick shellac varnish, letting it soak in, and finally bake it overnight in a warm oven to set the shellac hard. No binding wires will be required for so small a machine, but they should never be omitted in larger ones, or in longer armatures.

Connections to commutator should next be made. Two wire ends will, of course, be fixed to each segment;

the latter and the wires being thoroughly cleaned and soldered together, using resin as a flux. The soldering must be thorough, but care must be taken that the heat does not affect the insulation. It must be admitted that it is very difficult to solder the connections in so small a machine, and the alternative is to fit small screws in the segments and thus fasten the armature wire ends.

The field magnet core should be insulated with brown paper, and bobbin ends (made in two halves, glued together in place) will make a good finish. The field magnet can be centered in the lathe and thus wound, being turned round by hand. The wire must afterwards he shellacked and baked in the same way as the armature.

All surplus insulation should be cut from the armature, when it can be fitted to bearings on a neatly planed baseboard, which is cut away for the reception of the machine. The field magnet is not screwed down until it is so fitted that the armature runs quite truly in the tunnel. The pulley and brush rocker may then be fixed, and connection to terminals and field exciting coil made.

The first baseboard should be mounted on a heavier board, to which it can be permanently fixed. The larger board can be screwed to the bench, etc., so that the dynamo is held rigidly during driving.

As a motor, running at its proper "full load," this same machine will take a current of about 4 or 5 amperes at 6 volts, requiring three large bichromate cells to run it satisfactorily.

CHAPTER VII.

Useful Data for Dynamo and Motor Builders and Users.

A VERY common way of measuring the capacity of small dynamos is in candle-power, but this is to be deprecated, and the purchaser of a small machine should always quote the number of *watts* he requires to be developed by the dynamo at full load. By full load should be understood that output which a machine can be expected to give for a lengthened period (one hour, at least) without unduly warming up.

The output of an electro-motor is measured in horse-power. The brake horse-power is the quantity here understood, and this in a fairly efficient small motor may be about 70% of the power supplied to the machine. In models the efficiency is very much less, and it is very difficult to estimate their output of power; about 50% may be taken as the maximum in most cases.

Small dynamos are very largely used for electric lighting. The amateur can readily calculate for himself the amount of light his dynamo is capable of producing, from the folowing data. Incandescent lamps of two types are sold, one of "ordinary" and the other of "high" efficiency. The latter cost more and have shorter life, but use a small current per candle-power. Lamps of ordinary efficiency, of good make, consume from 3.5 to 4 watts per candle-power; average, 3.75. Thus the usual 16 c. p. 100-volt lamp requires a current of .6 ampère. High-efficiency (H.E.) lamps consume from 2.5 to 3 watts per candle-power, and it is advisable to use them in small installations. Lamps can easily be obtained of the voltages adopted for the various dynamos described in these pages.

A very few words on the subject of the amount of light required for any particular purpose will not be out of place. To light a room of about 8 ft. × 10 ft. at all satisfactorily, a 16-c.p. lamp will be required. A room 10 ft. × 12 ft., which will also probably be higher, requires two such lamps or their equivalent; while a bedroom is usually sufficiently well lit by a single 8-c.p. lamp. For a passage-way, pantry, or cellar, 5-c.p. is ample, although more must be provided in either case if a strong light is desirable. Owing to the nature of electric light, lamps can often be so advantageously placed that half the c.p. afforded by the ordinary gas jet may be found sufficient. In living-rooms it is wise to give an ample margin, the rates specified above for them being the minimum generally allowable.

The power required to drive a dynamo depends as much upon its efficiency as upon its output. One horse-power is equal to 746 watts, but a dynamo of 746 watts output requires much more than 1 H.P. to drive it in order to overcome the friction of bearings and brushes, besides a proportion of electrical energy which never appears in the outer circuit, being used partly to excite the field magnets and being partly wasted. All these have to be paid for in the form of extra power supplied by the engine; and in the case instanced probably not less than 1¼ actual B.H.P. would be required. The source of this power may be steam, gas, oil, or even hot-air engine, or water-motor; as long as sufficient *power* is supplied, the nature of its source is unimportant. It is advisable when using gas or oil engines to have extra-heavy flywheels fitted, or better still, to provide a fly-wheel at each end of crankshaft, so as to ensure even running.

A dynamo should not be driven by a steam engine a great deal too powerful for the work. Very small machines of fair make require about three times the power they nominally should; those between 200 and 400 watts about twice, and from 500 to 750 watts one and a quarter to one and a half times as much power is needed, as the electrical output would indicate. Owners of small dynamos or motors should remember that they are mechanical as well as electrical machines, and therefore require treatment in accordance. The bearings should be correctly designed and efficiently lubricated. Commutators should be quite clean, and generally require no lubrication in small machines. Belts or driving bands must be of even thickness, and the joints as good as possible, or lamps lighted from a dynamo will flicker every time the joint of the belt passes over the pulley.

"Direct" driving is not an easy matter with small machines.

When charging an accumulator, the dynamo must first run up to speed, and the voltage noted either by means of a voltmeter or pilot lamp. Only when the full voltage is attained may the cells be switched in. The charging current must be at a higher voltage than that of all the cells which are being charged in series, not less than 2.5 volts being allowed for each cell. Where a dynamo is used for lighting in conjunction with storage cells, being sometimes used to light the lamps direct and sometimes to charge the accumulators, it must be run at higher speed for the latter if all the cells are to be charged in series. Care should be taken when this is to be done, that none of the ordinary lamps can be in the circuit, or they will probably be burnt out.

The following particulars govern the use of small motors, such as may be built from the details contained in these pages: The proper voltage of the supply current to run any machine as a motor is practically that which is quoted as the voltage of· its output as a dynamo. To give an example, a 250-watt dynamo rated at 5 ampères 50, volts should be supplied from a 50-volt main. When running under full load its consumption of current will be about 5 ampères *plus* the amount required to energise the field magnets, and its output, if of efficient design, somewhere about a ¼ H.P. To explain a point which sometimes puzzles the student, it should be said that the current a motor will absorb when running is by no means that which a mere consideration of its internal resistance and the voltage alone might imply. Supposing the total resistance from terminal to terminal in the case of the above motor were 2 ohms and the armature held fast, it is true that a current of 25 ampères* would be forced through the motor and would quickly ruin it. The armature being left free to revolve, however, imposes a "back E.M.F." or pressure in the opposite direction to the supply current, acting virtually as a dynamo. If the motor is overloaded and the speed of the armature thereby reduced, a larger current will pass, which will to a certain extent enable the motor to take the extra load; it is indeed just this characteristic which makes electro-motors so accommodating. To prevent accidents through excessive current, fuses are placed in the motor circuit to prevent too great a current passing through the machine, and it is usual to check the great rush of current on first closing the circuit by interposing a "starting resistance," which is gradually cut out as the motor gathers speed. With small motors operated from primary batteries fuses at all events are unnecessary, as the battery cannot supply a very great surplus current for longer than a few seconds. A starting resistance, on the other hand, is of use if the motor is required to start without a violent jerk.

A small motor can be run from a *continuous* current supply main of higher voltage than the motor is

---

* Calculated from the formula, $C = \dfrac{E}{R}$. In the present case $E = 50$ volts; $R = 2$ ohms. Therefore, $C = \dfrac{50}{2} = 25$ ampères.

built to run at, if a suitable resistance is inserted in the circuit. This resistance should generally be a lamp of the proper voltage of the supply. The approximate correct current for the motor should be known, so that a lamp may be chosen which will pass sufficient, but not too much current. A 16-c.p. lamp takes .6 ampère on a 100 volt circuit, .3 ampère on a 200 volt circuit, and so in proportion, and for lamps of other candle powers in proportion similarly. Resistances for the above purpose may also consist of iron or special resistance wires, proportioned both for carrying capacity and for the amount of resistance required in the circuit. There is, of course, considerable waste, and the proper way to manage, wherever possible, is to build the motor with suitable windings to run from the main direct; this enables the most economical use of current to be made.

The following table of speeds shows the rates at which armatures of different sizes should be driven to

Diameter of Armature, Inches.	Speed-Revolutions per minute.	Diameter of Armature, Inches.	Speed-Revolutions per minute.	Diameter of Armature, Inches.	Speed-Revolutions per minute.
1	4000	2⅛	2750	3½	2250
1⅛	4000	2¼	2700	3¾	2200
1¼	3800	2⅜	2650	4	2150
1⅜	3500	2½	2550	4¼	2100
1½	3300	2⅝	2500	4½	2000
1⅝	3100	2¾	2450	4¾	2000
1¾	3000	2⅞	2400	5	1900
1⅞	2900	3	2400	5½	1800
2	2800	3¼	2300	6	1750

get the outputs quoted for them. It may be said that these speeds are approximate only, and may be altered if the machine is found to need it for a particular voltage. The speeds of motors may be taken as about ¾ of the speeds given for similarly dimensioned dynamos.

CHAPTER VIII.

HINTS ON TESTING AND REPAIRING.

THERE is nothing more difficult than to give a satisfactory answer to the not infrequent question—"I have a dynamo which will not work. Can you tell me what is wrong with it?" A question only to be answered thoroughly by a compendious reply, which may possibly be useless after all. It is intended in this chapter to enumerate all the most likely failings the dynamo (and motor) is heir to, but it must be assumed the reader is conversant with at least the elementary laws of electricity, and is posessed of a little of that instinct which is essential in the practical electrician. To simplify matters, the writer will take the various complaints of model dynamos and motors in the order of frequency as he has observed them.

"I cannot get my dynamo to work at any sped, although it runs well enough as a motor from—so many —cells," is perhaps most often heard. Sometimes the maker adds the information that he has tested it in every way, and is sure all is right, winding and connections correct, etc. The most probable reason why the machine will not act *as a dynamo* in this case is that the residual magnetism in the field magnets is of wrong polarity. The remedy is simple. Reverse the direction of the exciting current through the field-magnet coils by reversing their connections to the brushes, or reverse the polarity by passing a current from a battery through the coils. If it does not affect the position of the brushes, it will be sufficient merely to run the machine in the contrary direction. As already indicated (see p. 8) the cause of a failure of this character may lie in the extreme softness of the iron of the field magnets. This is, however, only likely to occur when charcoal iron stamping are employed.

If the fault is not to be found as above, the winding should next be considered. Small dynamos, especially with field magnets cast from unusual qualities of iron, may refuse to work with a field-magnet winding of "average" resistance. Indeed it is a good plan not to finish off the field-magnet winding until some experiments have been made to determine the quantity of wire which will give the best result. Mistakes sometimes occur in the gauge of wire supplied, and an error of this kind would almost certainly be fatal to the efficiency of the machine. The connections of the armature windings to commutator should be examined, in case any mistake has been made, and to see that the brushes rest on the bars connected with wires conveying the maximum current. If a dynamo refuses to "build up," the brushes may be given a slight backward lead, when the armature tends to produce its own field. As soon as the machine starts working, however, the brushes must be advanced, and in no case should they be moved far from the neutral line. A number of other reasons depending on the small amount of magnetism retained by soft metal may be mentioned. A slightly wide air-gap—more than the average, but still not excessive—may be quite sufficient to account for the failure; the weak magnetic flux is unable to bridge the air-gap sufficiently to produce an appreciable current in the armature wires, and if these happen to be too thickly insulated, so that the influence of the iron core of the armature is diminished, it will be quite enough to account for the refusal of even a fairly well-designed generator to develop current. It should be remembered that when the machine is running as a motor, supplied with an ample current, the magnetic lines of force are numerous enough to overcome wide air spaces and thick insulation, and to render slightly disproportionate windings of little account.

A method of overcoming this defect has been given to the writer by an electrical engineer who has had great experience in designing and working small dynamos. It is to wind soft iron wire of any suitable gauge from end to end of the armature so as to completely cover it, securing the ends of the wire by soldering or otherwise.

One entire class of failures is due to the breakage of a conductor, either in the field-magnet coils, armature winding, or connections. The effect is refusal on the part of the dynamo to generate at all, or to cause a pulsating or intermittent light when the machine is supplying current to lamps. In the case of a motor, it may entirely refuse to work or may work with varying and irregular speed. Fortunately the location of a fault of this character is not a very difficult matter, and involves only the use of a battery of any ordinary kind and a small galvanometer such as can be purchased for about 25c. A battery of *small* cells, and many in number, will be found most useful for testing, especially if the dynamo is built for any fairly high voltage.

The field-magnet winding should first be tested by disconnecting the ends of the coil, connecting one end to one terminal of galvanometer, the other terminal of galvanometer to one electrode of the battery, and then joining the free end of field coil to remaining electrode of battery. If all is well, the galvanometer needle will be deflected. If no deflection occurs, proving that a wire in the circuit is broken, the separate coils of field-magnet should be each tested if there is more than one.

A failure in the armature is a more troublesome matetr both to test for and to repair. With armatures of the "open-coil" type the test consists in making contact through galvanometer, battery, and two commutator segments, all the segments being tried in turn. With a broken wire, as before, no deflection will be observed. The armature connections to commutator bars should be particularly well examined; these are fruitful sources of trouble. The only way to test for a broken wire in the armature coils is to disconnect (unsolder) the junctions both between one coil and the next and between the wires and commutator. Then each coil can be tested separately. It is useless to test from one commutator segment to another for a break in a wire on a "closed coil" armature (drum and ring types), since the current has two routes open to it, and the breakage of a wire does not appreciably affect its flow.

Occasionally a dynamo is found to supply a fluctuating instead of a steady current, or a motor will run in a jerky fashion. This is usually traceable to a partial break, where the two ends of the wires are not entirely separated but scrape together when the machine is running. A test can be made for this defect with battery and galvanometer, by striking any part of the carcase with a light hammer, which will cause the imperfectly joined wires to jar and set up oscillations in the galvanometer needle. Tests must be made to locate the fault exactly.

A very troublesome class of failures is that in which the symtoms are evident when the machine is running but cease when the armature is still. Tests usually show in these cases a leakage from armature coils to the frame of machine, and it will be found that the cause is centrifugal force and a slack or loose turn of the armature wire. This latter is bulged all the time the dynamo is running and makes contact with the poles of field magnet. Possibly, however, the bearings are to blame, and these should be examined if the armature is running not quite centrally. If they are too slack, the fact will be indicated by noise as well as by considerable vibration.

Violent sparking at the commutator which cannot be reduced by altering the lead of the brushes may be due either to overloading or to a short circuit in the armature. It is easy to prove whether the former is the case, and if this is not the cause the fault must be attributed to the second possibility. Under the head of "short circuit," however, must be placed (in amateur-made machines) unequal numbers of turns of wire in each coil of the armature. This will produce much the same result, and as sparking of small machines is often somewhat excessive for their size, it must not be too readily assumed that the more serious fault exists.

It should always be remembered that a shunt-wound dynamo will refuse to "build up" with a very low resistatnce in the outer circuit. Thus, if a few cells of a new (uncharged) accumulator form the external circuit, a small lamp should be inserted at first in this circuit, or the electrician may be surprised to find no trace of current in the cells after an hour's steady "charging."

It is unnecessary to emphasise the fact that non-adherence to any of the rules or methods laid down in this book may lead to failure. If an armature is being wound it must be wound according to instructions, or it may prove useless.

Leakage of current is a very common trouble with small dynamos. If well made in the first place, the insulation being thorough and wires firmly bound where necessary, each part being tested with galvanometer and battery before fitting together, there is no likelihood of current leakage unless the machine meets with an accident, or is overloaded. Should the trouble arise, it must be tested for by means of a battery and galvanometer in the following way: First lift the brushes clear of commutator and disconnect the brush leads from terminal board. Then test each brush holder to rocker or bearing. If no sign of leakage is to be found here, test between commutator and shaft, and then between field winding and frame. If the fault is found in the brush holders and rocker (a very likely place) it can soon be set right by taking the holders to pieces, cleaning thoroughly, and making sure—when putting together again—that the brush holders are efficiently insulated from the rocker arm. In the event of the fault appearing in the field-magnet winding, a test should be made to find which coil (if more than one) is responsible, by disconnecting the exciting coils from one another. The faulty coil will have to be rewound, but it will be well to observe first that the trouble is not due simply to a loose turn or a bared connecting wire in contact with with the carcase of the machine.

The most serious case is when the armature winding is found to leak to the armature core. The same observation should first be made to see that no outer turn of wire is at fault. The writer has also come across a bad case of leakage (in a small motor) which was found to be caused by the fine binding wires employed to retain the armature coils in place. The maker in this instance had simply wound on a few turns of fine brass wire (bare) very tightly, without placing any insulation between them and the armature wires. The fine wire evidently cut through the cotton covering of the coils and established a too familiar acquaintance with the conductor wires inside; the cause was only discovered by making a test after the binding wires had been removed and before proceeding to unwind the armature

If the latter has to be done, the connections to commutator must first be broken, and the commutator then tested, each section separately, with the shaft. If it is still found necessary to attack the armature winding, begin by untwisting all the connections from coil to coil, and test each coil separately; this method, and, in fact, the method of testing step by step methodically, is certain to save time and ensure finding the cause or causes of failure. It should not be forgotten also, that the finding of one fault by no means precludes the possibility of another occurring in the same machine; and if a leak is found at all, all parts of the dynamo should still be tested, even if the very first attempt locates a fault.

A large number of dynamo and motor failures are due to excessive current. A small dynamo is very easily, and therefore very often, overloaded. Even its normal limit of temperature is high, implying a heavy current in the armature, and when this is doubled, the natural result is a rise of temperature in the lower coils of the armature sufficient to char the insulation. With small motors the case is even worse. The rush of current from a fairly powerful battery through a motor at first starting is very great, but this ceases as soon as the machine has picked up its speed and opposed its back E.M.F. to the supply current. When, however, so great a load is put on the motor that its speed is brought very low for any considerable time, it is likely to heat up, generally with disastrous results.

It is very important to have the brushes well pressed

on the commutator, but not heavily enough to create undue friction. To get satisfactory results the shaft should be a good "running fit" in the bearings, not slack; the commutator should be true and round, with no flat places or grooves, and the armature truly balanced. By these means, "sweet" running is assured, and this is essential, as the consequence of slackness in the bearings, etc., is a loss of current, excessive sparking, worn places on the commutator and noise.

Heating sometimes takes place in the field magnets. It is generally due to "eddy" currents set up by the cogs or slots in armature being few in number and rather wide. The only remedy is to overwind the armature with soft iron wire as described on page 65; but as prevention is better than cure, as large a number of cogs should be employed as possible.

We have now enumerated practically all the failures usually met with in a dynamo or motor which is not inherently useless by reason of design or construction. A dynamo of excellent constitution may "go wrong," and the foregoing hints are intended to show what should be done to test and to remedy it. If, however, it is a machine which always works badly, or can never be made to give current at all, the case is more serious, and only a practical examination will quite satisfactorily solve the problem. The windings may be entirely disproportionate or the design, i.e., the quantities and distribution of the iron, bad, and the reader whose dynamo is a failure should compare the machine with those described in this book, when he will doubtless discover the whole cause of the trouble.

To indicate the difficulty under which the expert adviser works when his experience is sought, a few of the cases which have come to the personal knowledge of the writer may be mentioned. There is first the instance—which has been more than once repeated—of an undertype dynamo very carefully mounted on an *iron* baseplate; as certain a means of short-circuiting the poles as a piece of wire across the terminals will short-circuit the current. Then there have been numerous cases of wrong winding of field magnets—resulting in two north or two south poles, as the case may be. This can easily be detected by means of the magnetic needle of a cheap galvanometer. Armature coils are often wrongly wound, the output being thus diminished or nullified, whilst the employment of plain armature stampings in small models, overwound with several layers of wire, is responsible for many more inefficient machines. In these cases, almost without exception, the makers were satisfied there was "really nothing wrong with their work," and laid the blame on the designers; but what is to be said of the budding electrician whose design and construction were excellent, his armature properly wound and connections faultless, yet whose machine would produce no current? He had had his field magnets cast and beautifully finished in the best gun-metal!

THE END.